Al

MOUNTAINSIDE

By Shannen Yauger and Kelley Williams

Illustrated by Tanya Glebova

Designed by Sara Re

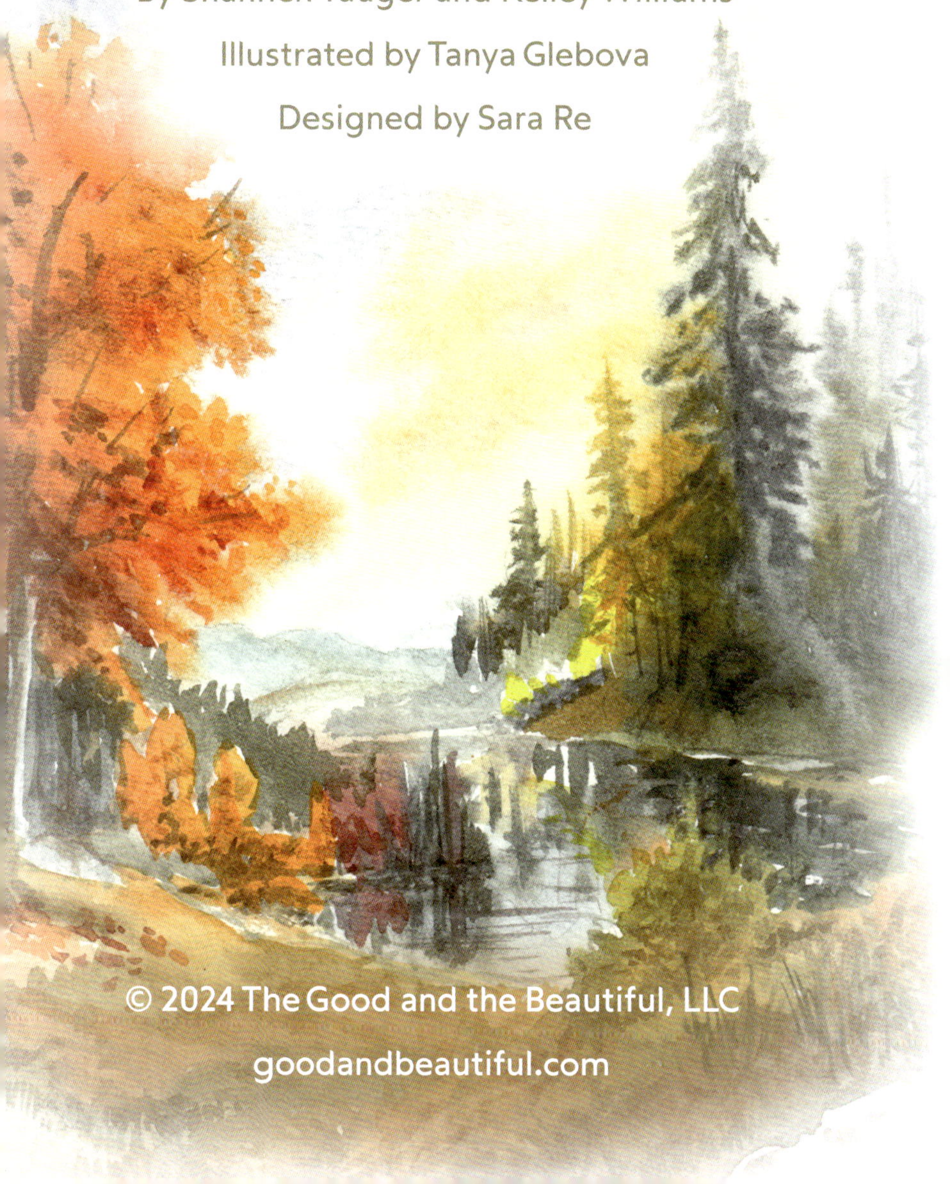

goodandbeautiful.com

Table of Contents

Winter 1

Spring 9

Summer...................... 17

Autumn 27

Winter
The Silent Snowfall

Along the base of the North Georgia mountains, winter arrives, a blanket of snow covering the forest floor. As the temperature drops and the days grow shorter, Red Fox prepares for the challenges that lie ahead. With her thick fur providing insulation against the biting cold, she ventures forth from her den in search of food. She relies on her keen hunting skills to navigate the chilly landscape; her long snout tracks along the forest floor, trailing the scent of a cottontail rabbit. Sleek black-tipped paws tread quietly on the freshly fallen snow as she works throughout the night

and into the early morning. As a golden sun rises over the mountains, her orange-red fur glimmers as it reflects the rising sun. She knows this mountain well, having lived here since she was a pup. Suddenly, she stops and turns her large ears to and fro. What is that she hears?

Not far from Red Fox, Wild Boar is foraging for acorns under the oak trees. Anything but quiet, Wild Boar sniffs and snuffles, his snout buried in the snow. Red Fox keeps a keen eye on Wild Boar's methodical

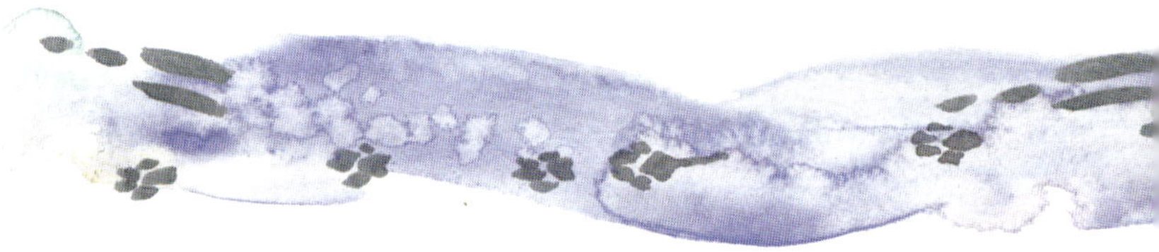

search for food. Intrigued, she follows him, keeping her distance as she watches him uncover hidden treasures on the frozen ground. Intent on his search for breakfast, the large animal presses along, snout to the ground, unaware that he's being watched. Acorns, roots, tubers, and even small mammals become his targets as he roams his territory. Despite his messy appearance, Wild Boar is a creature of routine and habit. He follows well-trodden paths through the woods, his thick hair providing protection from the biting cold.

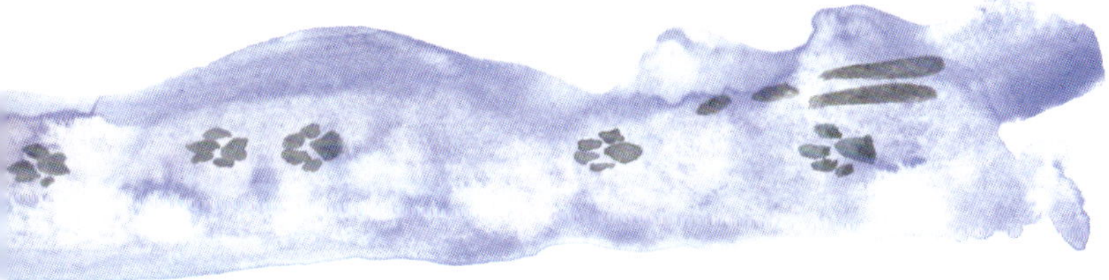

Pygmy Shrew, startled by Wild Boar's foraging, quickly scampers into hiding. This tiny animal is easily surprised, and the sudden appearance of Wild Boar's foot is an unwelcome interruption to his scavenging of leaves for insects and worms. Constantly hungry, Pygmy Shrew captures his prey often, eating a meal every fifteen to thirty minutes. His tiny body is in near-constant motion, searching the forest floor for food.

Food and movement are key to Pygmy Shrew's survival in the cold winter months, and he requires little sleep. Quickly, he scurries under the leaves, far away from the risk of being crushed by Wild Boar. He finds a few small insects to devour, though he will also munch on seeds or fruits when no insects can be found. Although his thick, dense fur works like a coat in the cold, Pygmy Shrew decides to seek shelter for a bit

of the day. Wet leaves rustle as he dashes away to his underground burrow, settling in to conserve his energy and avoid the harsh, cold wind.

No longer hungry after foraging for his meal, Wild Boar seeks shelter in dense thickets, finding respite

from the chill winds that sweep down from the mountains. He will rest here until dusk, as he is normally most active at night. Tucked away in the deep underbrush, he settles in to sleep.

Red Fox silently observes her den, where her mate is nestled.

The soft snow shows no trace of her passing, as her furry paws are like snowshoes, allowing her to walk on

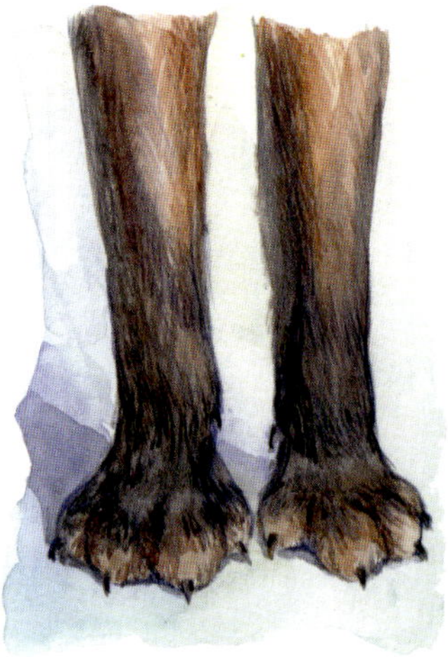

top of the fallen snow without sinking as she makes her way to her den with her meal. She must eat enough to sustain herself and the baby foxes she is carrying, for she will spend the rest of winter pregnant with new life.

Little does Red Fox know, she is crossing over the home of another animal that lies curled in a burrow, unaware of what's happening above. Woodchuck has been in a deep sleep since late fall, waking only once or twice to

take a short walk. Soon she will awaken and rejoin the other mountain creatures in their daily activities.

The forest is quiet now, save for the occasional rustle of leaves, the gentle trickle of a nearby waterfall, and the soft crunch of snow underfoot. The animals of the forest await the arrival of spring with eager anticipation, knowing that warmer days and plentiful food will soon return to the mountain wilderness. For now, they rest and rejuvenate, and as the last traces of winter fade away, the forest awakens to the promise of new life and growth.

Spring

A Time of Renewal

Red Fox emerges from her den, her coat now sleek and shiny as she basks in the warm rays of the late-afternoon sun. She spent the last few weeks of winter preparing a cozy space in her den to have her litter of fox kits. Spring has arrived, and with it, her kits. Now that they're here, blind, helpless, and covered in a soft gray fur, Red Fox's mate must find food for all of them. He trots away, bushy tail swishing from side to side,

to forage for mice or grasshoppers in the surrounding woods for his babies. Stretching in the sun one last time, Red Fox heads back into her den for the evening to keep the kits warm.

Sniffing along a fallen tree after sundown, Red Fox's mate has detected, with his keen hearing, the sound of a mouse digging under him. He crouches behind a large root, waiting for the mouse to make its appearance. The unsuspecting mouse pops out of his hole, and the fox pounces in one giant leap. Proud to have caught the next meal for his family, he runs back to the den with the small rodent in his mouth.

Behind a nearby pine tree, grunting and squealing, four wild boar piglets fight each other for the best nursing spot on their wild boar mama. The two-week-old babies are a jumble of fuzzy brown-and-cream stripes and tiny waving tails as they clamber over each other to feed. Soon they'll venture out of the safe thicket made of leaves and grass to learn how to forage and protect themselves in the forest with the help of their sounder—a large group of female wild boars and their offspring that stick together.

Wild Boar, the father of the piglets, has left the mother and babies to once again lead his solitary life, constantly on the hunt for his favorite spring food: young, tender grass. His bristly, thick winter coat is shedding, and his skin feels itchy from the loose hair and the ticks that have attached themselves to him. Rubbing against the rough bark of a nearby tree, to scratch his back and sides and remove the pests, provides some relief. Feeling better, Wild Boar

trots off to fill his hungry stomach, easily jumping over a wide stream. Catching the scent of a fish in the water, Wild Boar sniffs along the water's edge, his flat snout twitching. Timing it just right—there! He opens his jaws and plucks a tasty fish out of the stream. Wild Boar will take his prize back to his shaded nest

of leaves, where he will enjoy his meal and settle in to sleep through the warmest part of the day.

The beautiful spring weather encourages Woodchuck to emerge from her burrow beneath the roots of the oak tree after months of hibernation. Blinking as her eyes adjust to the soft sunlight, she stretches her tight limbs. Her thick fur, once matted and dull from lack of activity, now shines brilliantly in the warmth of the spring sun. She sniffs the air, detecting the crisp scent of pine. The last five months have been spent in a deep sleep, her heart beating only four or five times per minute. After burning through the fat stores from all the food she ate in the summer and fall, Woodchuck is ready to forage.

Before she can leave her winter burrow, a male woodchuck wanders by, greeting all the females in

the area as he hopes to be the father of a litter of woodchuck cubs in a few months. As he passes, Woodchuck scans the sky for any dangerous hawks or other predators. Deciding she is safe, she cautiously walks a short way from her burrow, nostrils flaring as she sniffs for her first spring meal. Her favorite blackberries haven't grown yet, but she finds some early spring tulips and begins to munch.

At the entrance to another animal's burrow, Pygmy Shrew stops, his long, narrow nose bending this way and that as he detects a spider nearby. Sneaking up on the unsuspecting spider, Pygmy Shrew sinks his teeth into the arachnid, releasing a venom that paralyzes the prey. His next meal taken

care of, Pygmy Shrew takes a few minutes to rest just outside the shade of the borrowed burrow, basking in the warmth of the spring sun before the upcoming rain.

After a wet month of April showers, the animals of the mountainside welcome the drier days of May. Scurrying along the fresh growth of grass and wildflowers, they enjoy the sunshine and new life that the season has brought and prepare for the heat of summer.

Summer
A Time of Abundance

June brings heat and humidity to the North Georgia mountains as the summer months blossom and bloom. Waiting for the heat of the day to pass, Red Fox and her little pack of kits roam the many tunnels of their shady den. Needing more space for the young ones to run, Red Fox cautiously heads out from the den to make sure the area is safe for her littles. Seeing no danger, Red Fox barks *wow-wow* to her offspring—a signal that it is safe to come out. The kits, who are now much bigger and busier than they were in the spring, spill out of the den entrance, wrestling with each other and playing chase. Every few minutes, a kit stops to drink nourishing milk from Red Fox.

Ears swiveling and nose sniffing, Red Fox picks up her head at the sound of her mate returning with a large fish for the family. Freshly caught in the cold mountain river that flows nearby, it is a tasty meal for the family, and the kits immediately hunker down and ambush the fish. They are practicing for the time when they will have to find their own food. Tearing at the fresh meat, each kit gets enough food to fill its belly for a while.

Red Fox is suddenly on alert. She has spotted a bald

eagle soaring overhead. With a long, shrill bark, she warns her babies to rush back to the safety of their den. The eagle circles and flies back over them, lower this time. One kit

is slower than the others. Determined to protect her young, Red Fox turns back, rears up on her hind feet, and bares her teeth at the eagle, growling and barking. The eagle gives up, and the baby fox safely makes it

into the dusty den to join its siblings.

The heat of the midday sun beats down on Mama Wild Boar and her piglets. They've spent the early morning rooting in a nearby field for juicy strawberries. With their bellies full, it's time to cool off. Unable to

sweat like other animals to cool their bodies down, the boars must head to their wallow—a muddy pit next to a stream. The piglets happily slide into the wallow nose first and roll around, coating their heated skin with cool mud. Mama Wild Boar follows them with a satisfied grunt, as the mud will also help remove the parasites that have made her skin their home. After a while, she calls to her piglets to leave the wallow and find somewhere to rub off the mud, as she is ready to take a nap until the sun sets and the air is cool again. One by one, the little boars trot after their mother.

A few miles away, Wild Boar wanders his territory, ears swiveling independently from each other as he listens for danger. He, too, is looking for a place to cool off. On his way to a small pond, he detects a snake slithering in the grass next to him. Snuffing along the ground with his leathery snout, Wild Boar scoops up the small snake to have for his meal. Satisfied with his meal, Wild Boar climbs over a large fallen tree and wades into the chilly pool near the stream for a swim. With just his nose, ears, and the top of his head above the water, he swims for the other side of the water, where he climbs out and searches for a patch of tall grass in which to hide and sleep the hot day away.

High-pitched squeaks sound from a thick patch of brush. Pygmy Shrew has encountered another of his species in his territory. Opening his small mouth wide, he bares his teeth at the intruder. Both shrews squeal and hiss as they tumble together, each trying to win the fight. Pygmy Shrew is slightly bigger and stronger, so he chases the visiting male shrew off with a final squeak. The fight has left him hungry once again, so he skitters off among the leaves in search of another meal.

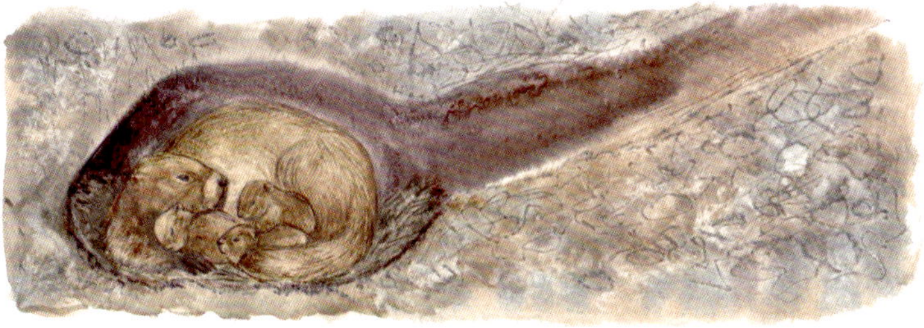

In the safety of her extensive burrow, where she has been raising a litter of three kits, Woodchuck is working away, digging more tunnels with her sharp claws. Three tiny noses come sniffing around the corner, and Woodchuck leads her babies the rest of the way to the compartment, where she has laid a fresh nest of leaves and grass. They are almost ready to leave the safety of her burrow to begin digging their own.

The woodchuck pups are hungry, so mom and babies venture out of one of the many burrow entrances. Hesitatingly, Woodchuck pops her head out and senses danger nearby—a coyote is slinking past

their burrow. With a warning whistle, Woodchuck ushers her litter back down the tunnel to wait for the coyote to move on to another part of the forest.

Once the area is safe, the woodchucks lope over to a nearby thatch of thorny blackberries. Sitting up on their haunches, the chunky rodents resemble tiny bears as they nibble on the juicy berries they're holding. When their stomachs are round and full, the woodchuck family returns to its burrow entrance to bask in the sun during the heat of the day. They will spend the rest of the summer eating, keeping their large, pearly white front teeth worn down to a healthy length, and working on their burrows as generations have done before them. By the time they're done digging, they will have a huge summer burrow in the field, full of winding tunnels and many rooms, each built for a different purpose, such as an area for sleeping, one for raising babies, and one for toileting.

The animals of the mountainside frolic on the forest floor as the heat of the summer begins to fade away. The leaves on the trees begin to change color and softly flutter to the ground.

Autumn

A Time of Harvest

After spending all summer exploring farther and farther away from their home and practicing their hunting skills, Red Fox's full-grown kits are ready to move out of the den come autumn. They have learned everything about surviving on their own from their mother. One of the kits begs Red Fox for food and milk, whining softly. Growling in return, Red Fox nips at the young fox, letting it know that she won't be giving the kits any more food. She

stretches out on the rust-brown and deep orange leaves carpeting the forest floor, waiting for nightfall so that she can go hunting for herself. Her mate has already left to spend the fall on his own, and she will spend the next couple of weeks encouraging her offspring to head out to find their own territories.

As the sun sets, Red Fox is on the prowl for food. Not only does she need to feed herself tonight, but she also needs to store food for the long, chilly winter in the mountains. It will be hard to find her favorite berries and other fruits once most of the plants go dormant. Crouching down in some tall

grass, ears swiveling, she hears the chirping of some nearby crickets. Several minutes later, Red Fox is busy bringing the crickets, one by one, to a hiding spot she has prepared under a tree. Using her nose to cover her catch with wet leaves, she heads off to find more food to store for the coming winter months.

With the mornings becoming cooler, Wild Boar can spend more of his day wandering his territory and hunting for food instead of hiding from the summer heat. As most of the fruits and vegetables have died off,

Wild Boar hunts around in the freshly fallen leaves for one of his favorite foods: acorns. His wide, rubbery snout pivots this way and that as he searches for a snack. After unearthing and chomping down several acorns, Wild Boar settles down in a thicket to rest. A hungry crow lands nearby and stands, watching as Wild Boar dozes since she knows there are likely some tasty parasites on Wild Boar's hide. Hopping closer, the crow tentatively pecks at a spot on Wild Boar's round stomach. Drowsily, he lifts his head to see what is touching him and then lies back down, uninterested.

Crouching behind a rock, heart racing faster than

any other mammal on the earth, Pygmy Shrew is stalking one of his preferred prey: a shiny red beetle. His whiskers twitch as he pounces, scoring another meal to fill his ever-hungry belly. After eating, he heads off through the underbrush, moving rapidly along the ground until he comes to the stream's edge. Leaning over, Pygmy Shrew begins to drink his fill. Overhead, an owl swoops low, searching for its own dinner. Most small animals in the forest would frantically run to a hiding place, but Pygmy Shrew isn't worried.

He will not taste good to the owl, as his back legs produce a terrible-smelling liquid when he is startled. The owl thinks better of his plans and flies off with a screech. Thirst taken care of, Pygmy Shrew sorts through some leaf litter on his constant search for more food.

Woodchuck has done her job of raising the next generation of woodchucks well, and the babies have left the nest. Her round body swaying back and forth as she walks, it is time for Mama to prepare for winter, and she heads out from her burrow in search of a filling meal. Nearby maple trees are a wonderful place to search for some tasty June bugs, and she begins to climb the closest tree. Halfway up, Woodchuck finds what she is looking for, and she balances on the crook of a branch to take bites of the protein-rich insects. Eating worms, grasshoppers, beetles, and other fatty foods will help Woodchuck build a layer of fat, which she will burn to sustain herself through her long winter sleep.

As the days and nights grow colder, the mammals of the mountainside ready themselves for another winter in the North Georgia mountains. A year of growing and learning has prepared them for the season ahead.